RIVER FRIEND

A series of Riverine Small Books

by Sylvia M. Haslam and Tina Bone

BOOK 4

INTERPRET:

What Do Plants Tell Us?

Book 4

INTERPRET: What Do Plants Tell Us?

A Book in a series of Riverine
publications by

Sylvia M. Haslam and Tina Bone

*Written and Edited by Sylvia Haslam and
Tina Bone. Illustrated by Tina Bone
(unless otherwise stated)*

RFS4: PAPERBACK 58pp.
ISBN No. 978 1 9162096 5 7
130 Illustrations

Published by: Tina Bone UK
First edition: June 2020

www.riverfriend.tinasfineart.uk
Email: ourbooks@tinasfineart.uk

CONTENTS

INTRODUCTION TO THE SERIES

Rivers are vital. They bring freshwater to the land, on which all its life depends. They are beautiful and fascinating, making up both the typical British countryside and many of its most spectacular views. If they vanished, what hardship and outrage there would be! Yet, slowly, slowly, they are vanishing, the larger stream becomes smaller, the tiny brook becomes a ditch and dries, and is filled in— the small ditches get polluted and dug out, become dull, and vanish from sight and consciousness. How can we save our rivers and riverscapes? How can we raise awareness on this slow, almost invisible loss?

We believe that this series of handy, small books, suitable for readers from teenage upwards, will help to raise awareness. Individually, each book tells a story on a particular riverine and riparian environment. Collectively, the series will inform, in a simple and effective manner, the invaluable worth of freshwater and its plants.

The Authors realised that there was a huge gap in the literature. There are many publications for scientists, for pond-dippers, birders and anglers, but "easy-read" books focussing on the river itself, and the vegetation belonging to it and creating the habitat for all else: we could find none!

For explanations regarding British freshwater plants, terminology mentioned throughout the series, and Picture Guide and reference section for further reading, see the book entitled *A PROLOGUE TO THE SERIES: Plant identification and Glossary of Terms* (also available to view free on-line at http://riverfriend.tinasfineart.uk/product/a-prologue-to-the-river-friend-series-isbn-978-1-9162096-2-6/)

Other titles in the Series are listed on the last page of this book and on the River Friend Website:
http://www.riverfriend.tinasfineart.uk

INTERPRET: What Do Plants Tell Us?

Leisure by **William Henry Davies**

What is this life, if, full of care,
We have no time to stand and stare.
No time to stand beneath the boughs,
and stare as long as sheep or cows.
No time to see, when woods we pass,
where squirrels hide their nuts in grass.
No time to see, in broad daylight, streams full of stars,
like skies at night.
No time to turn at Beauty's glance,
and watch her feet, how they can dance.
No time to wait till her mouth can,
enrich that smile her eyes began.
A poor life this if, full of care,
we have no time to stand and stare.

Fig. 1. A beautiful "Life" poem (*Leisure*) by William Henry Davies (1871–1940). We need to be able truly to stand and stare, and look and see, when it comes to river habitats and the water plants that live in them. (This poem, in the public domain, is the essence of this little book.)

Introduction

This Series of little books is about flowing waters, their vegetation, history, pollution and other interesting riparian/riverine aspects. Each book addresses a separate subject in its small space, but cannot, unfortunately, discuss the whole of that subject. Just think of all those places and things which could possibly pollute a stream and how they might affect all parts of each river plant, let alone harm all the other river organisms! Each book in the River Friend Series therefore has to be just an introduction to its subject. If this book (and others in the series) presses your "wish-to-learn-more" buttons, we hope you will delve deeper and explore and expand your knowledge via higher level literature.

This book, like others dealing with plant names, may at first glance appear to be rather intimidating, but this is only because even the English let alone the *Latin* names of the plants are unfamiliar. Think of your garden in early spring. There are some nice new leaves, long (10–12cm maybe), thin leaves under this tree and over in that grassy corner. They are in little groups, slightly blue-green in colour. In a few weeks' time they will be double or treble the height, maybe up to your knees. You have named the plant even before you have read this far. **Daffodils**! And if there has been any doubt about the leaves being the right size, that is removed when golden trumpets appear—the harbingers of spring—definitely Daffodils! Who cares that they may be hybrids scientifically named as *Narcissus* × *incomparabilis* (Fig. 2),

Fig. 2. The Common Daffodil
(*Narcissus* × *incomparabilis*)

and described in that community as: "Leaves 8–15mm wide, broad, glaucous, flower solitary. Perianth 4–6cm across, pale yellow, corona 5–12mm, deep yellow, cup-shaped, about half as long as perianth F1. 4. 2n=14, 21, GB. (Clapham, Tutin & Warburg, 1952)." Although some might remember "Nonesuch Daffodil"—one of its many common names—it is just a very pretty plant to most of us!

Who knows or cares? Well, botanists not knowing daffodils *do* care. The rest of us rarely need to. But this recognition by looks is how to get the best interpretation. "Oh, there is *Apium nodiflorum* [Fool's water-cress]. Look how small…!"

Excluding plants of the bog and moor (peaty) streams, there are only about 70 British river plant species which are used for diagnostic purposes. Most of these are listed and pictured in the "Picture Field Guide to some Common British Aquatic Plants" section in the book in this series entitled: *A PROLOGUE TO THE SERIES: Plant Identification and Glossary of Terms*. Whilst, of course, there are *more* species, if the 70 diagnostic plants can be recognized, the (non-bog) stream can "speak". If this little book encourages you to look more closely, try to identify and name one new plant mentioned here each time you visit a stream or a river. Get to know what the plants and vegetation look like when looking down at the water. A few very common aquatic plants are shown in Figure 3. All the plant species mentioned in this small book are illustrated at least once, but may also be repeated within context for clarity.

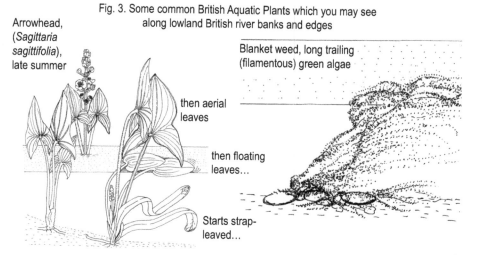

Fig. 3. Some common British Aquatic Plants which you may see along lowland British river banks and edges

Arrowhead, (*Sagittaria sagittifolia*), late summer

then aerial leaves

then floating leaves…

Starts strap-leaved…

Blanket weed, long trailing (filamentous) green algae

3

Fig. 3. Continued.../

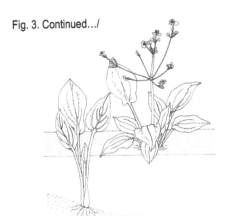

Water-plantain (*Alisma plantago-aquatica*)

Bulrush (*Typha latifolia*)

Bur-reed (*Sparganium erectum*)

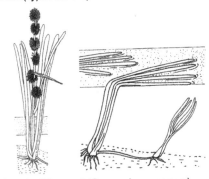

Unbranched Bur-reed (*Sparganium emersum*)

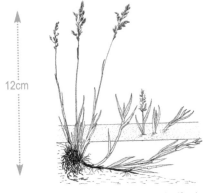

Creeping Bent Grass (*Agrostis stolonifera*)

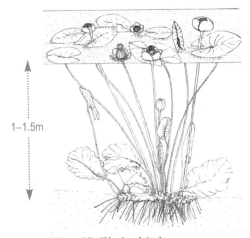

Yellow Water Lily (*Nuphar lutea*)

4

Fig. 3. Continued…/

Species of Duckweed
They have floating "leaves" (thalli) with short roots in the water below. The Common one is *Lemna minor/gibba*, with 1.5–4mm thalli and one root. Greater Duckweed (*Spirodela polyrhiza*) is 5–8(–10)mm, with several roots. Ivy-leaved Duckweed (*Lemna trisulca*) is submerged (just under the surface, see Fig. 8a) and translucent. The much smaller *Lemna minuta* is invading Britain from east to west

Marsh Marigold or Kingcup
(*Caltha palustris*)

Get to know the water plants as you know and understand "daffodils", not as you read the scientific "*Narcissus* ✗ *incomparabilis*" description above.

Before moving on to how the plants are speaking to us, there are further obvious points which are seldom put into words. If you have a dog it will be small when born: it must be. Then, all being well, it grows to the size appropriate for its breed (Cairn, Airedale, etc.) and it then stops growing. Children do the same. Illness or malnutrition can stop an animal growing to its proper size, but it is very difficult to make an animal grow beyond its "normal" full size. Hence you will never see a really old dog the size of a house!

Plants, on the other hand, normally grow more each year, though different parts grow differently. When an oak tree reaches the proper size for an oak tree, every year its trunk still expands and it still grows new stems and roots which grow indefinitely, and also new leaves and flowers, each having definite growth. With all this growth, the new and old mosaic of habitats on the tree increase too. More and more leaves grow but they, unlike the roots, are of roughly fixed size. When you next go walking, look at them—they are similar in size and shape.

With plants in general, better conditions—water, nutrients, light—mean better growth and better health. For the dog, good conditions also mean good health but not much more growth-bulk of the dog.

To get used to plants, individually and as vegetation, they must be learnt; to be learnt, they must be seen; to be seen, they must be "stared" at.[*]

Fig. 3a. Yellow Flag and Pond-sedge

---

[*] *A note from Sylvia Haslam:* I did not learn only by standing and staring, but also by recording what I saw (walking from car to river, usually from a bridge)—more properly recording (at first) the width, flow, and other basic features of the river, the plants present, and anything that seemed odd. It was not until after two summers of field research (two recording seasons) that I had an understanding of what I was seeing. My research had come to rivers from wetlands, and for those first two years I used to say that if I walked into a wetland I knew what I was looking at; if I looked at a river, I did not. Understanding then slowly grew. And as I published what I learnt, others—though still needing to "stop and stare"—were able to learn quicker than I did.

Now over to the plants!

Fig. 4.

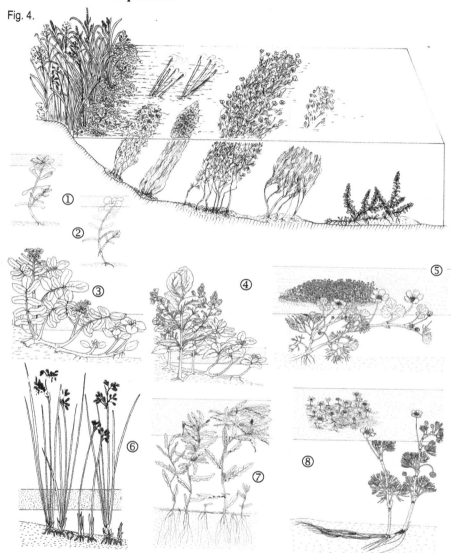

Figure 4: 1-2. *Callitriche* spp. (Water Starwort). 3. *Nasturtium officinale* (Water Cress) 4. *Veronica beccabunga* (Brook Lime). 5. *Ranunculus aquatilis* (Common Water crow-foot). 6. *Scirpus lacustris* (Common Clubrush). 7. *Potamogeton crispus* (Curled Pondweed). 8. *Ranunculus trichophyllus* (Threadleaf Water Crow-foot).

Why use Plant and River Indices (Classification)?

The stream or river is a continuum, reaching from its source right down to its mouth, which may be another freshwater-way or the sea. This may be a distance of only one or two kilometres or many hundreds of kilometres and the amount of vegetation varies along its length. To help researchers, and to standardise the literature, stream sizes and vegetation lists are normally categorised and these are used to describe portions of a stream or river which have something in common. Table 1 shows one proven Classification Model—invaluable as a scientific analytical tool. Within this model, the plants may be described as "None (absent), Poor, Fair, Good, Very Good, Exceptional".

Table 1. Stream Size Classification

Size (i)   Small streams (brooks or ditches) without water-supported species (*plants supported by the water, floating, and submerged species*), up to 3m wide. Emergents (*a plant mainly or entirely above water*) can be present or absent.

Size (ii)   Small streams (brooks) with **deeper** water than (i) and with water-supported species, up to 3m wide. Water-supported species present, emergents present or absent.

Size (iii)   Medium streams (small rivers), 4–8m wide. Any type of vegetation (*along the edges, or further into the water*) or empty.

Size (iv)   Large streams (medium and large rivers), between 10–30m wide. Any type of vegetation or empty.

Wetland Channels usually have negligible flow, for example, dykes and drains, and are graded similarly to the streams and rivers. Strangely enough, many of these channels have aquatic vegetation in abundance, and even hold rare species.

There are several scientific River Survey Classification Methods available, carried out for instance by the Environment Agency (System of Evaluating Rivers for Conservation Value—SERCON), by the Royal Society for the Protection of Birds (RSPB); and by various Wildlife and Countryside Trusts. Each method was designed for a particular purpose and all of them cover natural river ecology.

There are two accepted Schemes of Classification for Vegetation which, necessarily, overlap. One uses species presence, with a key, and the other uses not only the characteristics of the plant communities present but also the physiographic habitat (geography of the land). Below are some examples of how a Classification System assists with diagnosis. It is fascinating how many deductions can be made by scrutiny of a site's vegetation and stream type.

In the mention of aquatic plants in this book, the English name is listed once with the *Latin* name in brackets. Thereafter, the *Latin* name is written.

Example 1: Medium Chalk Stream (and/or Limestone)

Fig. 5. Recently restored Medium Chalk stream (July 2018): Mill River, Wendy, Cambs.
In Britain, Chalk streams occur from Dorset to Kent and from Hampshire to Yorkshire with a few in and around Cambridgeshire

Unlike animals, plants have a greater range of ways by which to respond to changing conditions, being able perhaps to adjust their number, size, shape and colour. New shoots and roots can all be varied more than can an animal. Many people do not notice water plants—it is just green vegetation at the riversides. But there is far more to it than that—if you "stand and stare" more deeply, just by looking at two plants of the same species you can spot many differences between them.

Each macrophyte (aquatic plant) species has its own "preferred" habitat, for example, a particular substrate on the bed (Fig. 6), fast water flow, slow water

9

flow, high nutrients ("inorganic fertiliser", calcium, phosphate, etc.), or low nutrient level. Many preference requirements overlap. A group of species may be characteristic of fast, or slow water flow, but none appear to be identical. Each species is different. This means plant communities "speak" to us: if just one species is present, the habitat may be anywhere in the spectrum of that species' habitat; if there are three present, the habitat has to be suitable for all three species.

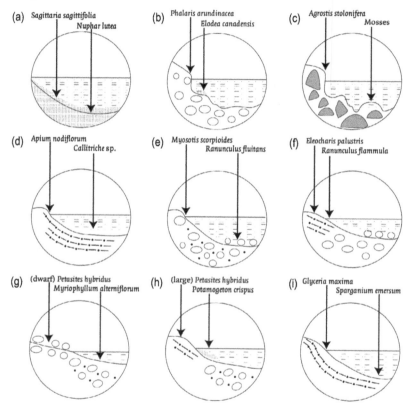

Fig. 6. Variation of species with substrate. The diagram shows a variety of substrates, from large boulders (circles) to silt (dots) (dash plus dot is "soil", mixed substrate), with typical species for each texture. The central line is that of the hard, or consolidated bed, with loose material lying above it. Nutrient status varies: (a) and (i) are high; (f) and (g) are low; and the rest are medium

Again, why use species lists and plant-indicator Classification? This is why: the site is 30m long, in a slightly silted/polluted (non-pristine) river. The average river vegetation cover is 75%, and the diverse species present probably include:

10

- **Tall monocotyledons** such as Grasses (*Glyceria maxima, Phalaris arundinacea*) and *Sparganium erectum*. A fringe of three frequent species.

- **Short emergents** such as Lesser water parsnip (*Berula erecta*), Floating grass (*Glyceria fluitans*), Water mint (*Mentha aquatica*), Monkey flower (*Mimulus guttatus* agg.), Water forget-me-not (*Myosotis scorpioides*), Water speedwell (*Veronica anagallis-aquatica* agg.), Brooklime (*Veronica beccabunga*). A mixed, dominant, frequent to abundant fringe of seven species, growing together.

- **Water-supported species** such as *Berula erecta*, frequent, submerged in a carpet; Water starwort (*Callitriche stagnalis*), occasional; Canadian pondweed (*Elodea canadensis*), occasional to locally frequent; Pondweeds (*Potamogeton crispus*), occasional, and *Potamogeton perfoliatus*, occasional; Water crowfoot (*Ranunculus* sp.), abundant. Six species in all.

...And the perceivable diagnosis:

The species list showing all these plants growing together in a 30m stretch of river would suggest that, scientifically, it is the middle reach of a lowland limestone river, probably 6–15m wide, with the water depth at its centre being mostly between 50cm and 75cm. The water is clear and of moderate flow. Also the banks are (mostly) not steep, and the edges are wide and shallow. The centre of the river bed is partly firm gravel, partly softer, with some silt at the sides. There is no recent major disturbance, shading by trees or buildings, or channelling, though there has been shallowing (drying) and perhaps a little grazing. Pollution is low and the water is chemically calcium influenced (calcium-rich, with other nutrients too low to fully reduce the calcium influence).

By studying species ecology, diagnosis can be much fuller. A site like this would be rated as "Very Good" whatever Classification Index was used, particularly when there are no pollution-favoured species such as Blanket weed present, and only the normal proportion of 4 out of 16 pollution-tolerant ones, such as *Sparganium erectum*. All three habit groups of tall monocotyledons, short emergents and water-supported species are represented. The highest number, both absolutely and relatively, is of the short emergents (6 bushy fringing herbs and one grass). Therefore there is enough space for this wide-edge habitat, and shallow water at the edge. Something is preventing the tall monocotyledons "invading" and shading out the short plants. This could be shallowing (exposing soil, which tall monocotyledons have not had time to colonise), unequal grazing, recent wash-out, unusual lining of the stream bed, fencing, or other factors (Fig. 7a–e).

Fig. 7a. Agricultural run-off ditch (into brook in valley) showing (left to right) overgrown at top of field; cleared but but not recently dredged mid-field; and recently dredged—run-off water has returned, under-drainage exit pipes in field have been cleared

Fig. 7b. Cattle disturbance, trampling and grazing

Fig. 7c. Stream Dredging

Fig. 7d. Cutting by hand

Fig. 7e. Man-made bed of unstable substrate (concrete, boulders and other loose, coarse material). Ex-ford replaced by bridge, now disintegrating, with few or no water plants

At this particular site there are only 6 water-supported species present—a small number in relation to the 9 emergents, which is a high proportion. This suggests that recent shallowing has occurred, with perhaps a little grazing. There are no rarer species present, such as Ivy-leaved duckweed (*Lemna*

12

trisulca, Fig. 8a) and Opposite-leaved pondweed (*Groenlandia densa*, Fig. 8b). This supports a diagnosis of within-water damage. Absence of species can only rarely be used as primary evidence, but it can be helpful in supporting a diagnosis. *Lemna trisulca* usually indicates calcium influence, *Groenlandia densa*, some richer (clay) influence.

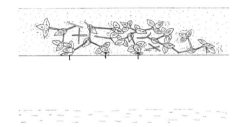

Fig. 8a. *Lemna trisulca*   Fig. 8b. *Groenlandia densa*

If we look at plant behaviour in even more detail, it can be seen that Reed sweet-grass (*Glyceria maxima* Fig. 9a) is associated with silt, so there is silt (at least at the sides) and it is not a fast-flow species. Reed canary-grass (*Phalaris arundinacea*, Fig. 9b) on the other hand, has a wider growing range which can be unflooded—at least in late summer. Both species can anchor to the bank: so, if stable, the substrate is at least partly earthen.

Fig. 9a Reed sweet-grass (*Glyceria maxima*)   Fig. 9b. Reed canary-grass (*Phalaris arundinacea*)

13

Sparganium erectum, being beside the bank in shallow water, is easier to dislodge in storms. Being so common, it merely diagnoses sufficient soil space and texture for its shallow rhizomes, deep roots, and infrequent wash-out. A good fringe of tall monocotyledons indicates firm banks, penetrable by roots and rhizomes, not greatly cropped, grazed or otherwise disturbed.

At this site there are 6 fringing herbs (bushy short emergents) forming a mixed and wide fringe. Mixed dominance comes with low nutrients which hinders growth and so the plants remain small and cannot grow into large mono-dominant clumps. As many as 6 such species indicates that they are growing in limestone, and because of the poor anchorage of plants in a mountain limestone stream, that this particular stream is most likely to be situated in lowlands (Fig. 10).

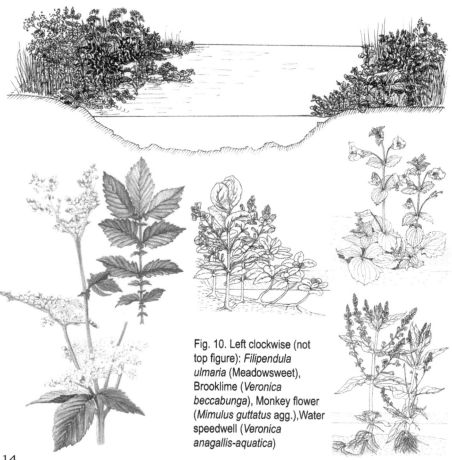

Fig. 10. Left clockwise (not top figure): *Filipendula ulmaria* (Meadowsweet), Brooklime (*Veronica beccabunga*), Monkey flower (*Mimulus guttatus* agg.),Water speedwell (*Veronica anagallis-aquatica*)

Amongst this group the Monkey flower (*Mimulus guttatus*) has the lowest nutrient range, so the site cannot be more nutrient-rich than a middle-sized limestone river. Water mint (*Mentha aquatica*, Fig. 11a) is slightly more nutrient-rich in range, being mainly in hills or on lowland limestone. Water forget-me-not (*Myosotis scorpioides*, Fig. 11b) requires the most nutrients. Within a range of Chalky (calcareous) lowland habitats (upper or middle reaches), all 6 species are interchangeable.

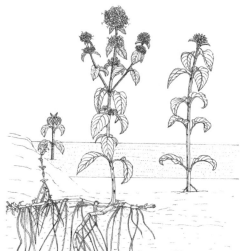

Fig. 11a. Water Mint (*Mentha aquatica*), can grow to a height of **80cm**

Fig. 11b. Water forget-me-not (*Myosotis scorpioides*), height up to—**50cm**

Berula erecta (Fig. 12) is not just in the emerged fringe, but is also present as a submerged carpet which indicates a limestone substrate, stable but active flow, and shallow water (up to about 50cm deep) which is usually clear. Pollution is always present in agricultural, settled, lowlands, but here it is mild and insufficient to overset the calcium-dominated influence of the water. As a submerged carpet, *Berula erecta* curls round superficial stable stones on gravel.

Water-crowfoot (*Ranunculus*) is usually in fairly clear and flowing water, although some species can thrive in slower or even still water. Other species, such as Common water-crowfoot (*Ranunculus peltatus/aquatilis*) can even grow in winter in shallow-flow limestone waters, only dying back in summer when the bed dries up. These variants are called "winterbourne ecotypes". At the other extreme, River water-crowfoot (*Ranunculus fluitans*) inhabits deep and often spatey rivers. Water-starwort species (*Callitriche obtusangula*

15

/stagnalis) are wide-ranging, usually growing in shallow water either at the edges or in the centre. This group has very dense, thin, shallow roots, and spreads rapidly by fragments. Wash-out from spates both removes and spreads the plants. Where it can anchor well, it can grow to dominate in shallow streams. This means there is both a suitable chemical as well as water environment, such as sandstone, with only moderate wash-out.

So, in small brooks, much *Ranunculus* and little *Callitriche*, means limestone; and much *Callitriche*, with little *Ranunculus,* means sandstone or altered limestone (Fig. 13a–e).

Perfoliate pondweed (*Potamogeton perfoliatus*, Fig. 14) needs adequate space in which to grow and a water depth of at least 75cm. This indicates a middle or downstream reach of river. The water is usually clear (clean or slightly polluted). Its substrate is soft enough for deeper roots to penetrate, yet firm enough to maintain anchorage.

Fig. 12. Above: *Berula erecta* showing emergent growth (silt edges) Top centre: *Berula erecta* sketch showing development growth. Top right: and (below) submerged carpet growth (limestone)

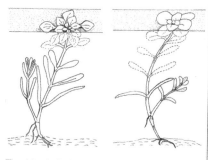

Fig. 13a. Left: Blunt-fruited water-starwort (*Callitriche obtusangula*). Right: Pond water-starwort (*Callitriche stagnalis*)

Fig. 13b. Centre to bottom across photograph, the submerged plant mass is Water Starwort (*Callitriche*) spp.. To the trained eye, top centre shows the rooting pattern of a nice submerged carpet of *Berula erecta/Apium nodiflorum*

Fig. 13c. Water-crowfoot (*Ranunculus*)

Fig. 13d. The photograph shows *Callitriche* spp. (top centre) and *Sparganium emersum* floating on the surface (centre), and emergent *Sparganium erectum* top and right, in a lowland, medium-flow, lime-base stream

Fig. 14. Perfoliate pondweed (*Potamogeton perfoliatus*)

Fig. 13e. Winter-growing (winterbourne) Water-crowfoot (and Curled pondweed, left centre) still growing quite vigorously in a little wild garden stream in February 2020. Cambridgeshire

17

When *Potamogeton perfoliatus* occurs with a carpet of *Berula erecta* this indicates that water depth is variable: *Berula erecta* grows in shallow water; *Potamogeton perfoliatus* grows in deeper water, so, for example, the two will be in deep and shallow portions of an undulating stream bed (pool-and-riffle system). There is probably little, or no, mechanical large-machine cutting or dredging. (Manual or small-machine operation seems to be less damaging.)

Curled pondweed (*Potamogeton crispus*, Fig. 15a) has deep, straight roots, longer than those of Canadian pondweed (*Elodea canadensis*, Fig. 15b). Both these plants anchor well in penetrable substrates. *Potamogeton crispus* may occur on silty edges of otherwise hard (or eroding) substrates, or in the centre of more mixed beds. It avoids strong calcium dominance, so in a Chalk stream it occurs downstream where a raised nutrient status (eutrophication), or incoming pollution, has lessened this.

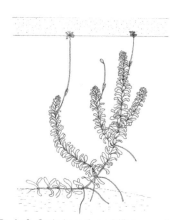

Fig. 15a, Left. Curled pondweed (*Potamogeton crispus*). Fig. 15b, above. Canadian pondweed (*Elodea canadensis*)

Deduction: By considering together the *Berula erecta* carpet (Fig. 15c), the presence of *Potamogeton perfoliatus*, *Elodea canadensis* and *Potamogeton crispus*, it can be deduced that there is lime influence, shallow water, with a firm gravel bed, as well as a deeper water habitat with a more penetrable bed.

Fig. 15c *Berula* Carpet

Example 2: Upland waters

Fig. 16

This example (Fig. 16) comes from a mountain river where, in the smaller streams, no 30m site had sufficient diversity or cover for the plants to be used for diagnosis. However, by amalgamating the data of ten sites, the resultant species list recorded can help. There was *Agrostis stolonifera* present in 5 sites, *Mentha aquatica*—2 sites, *Mimulus guttatus*—4 sites, *Petasites hybridus* (not large)—4 sites, *Phalaris arundinacea*—3 sites, *Veronica beccabunga*—2 sites, Mosses—8 sites, Blanket weed—1 site and the sole species. (Blanket weed (Fig. 17) is pollution-favoured, and this single site has no other species present; it is moderately to severely polluted, as diagnosed from the lack of other vegetation. Being just the one site, it should be separated from the others.)

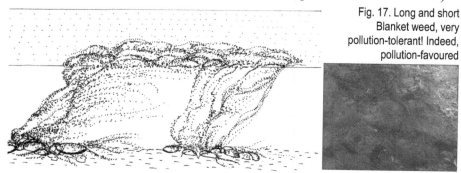

Fig. 17. Long and short Blanket weed, very pollution-tolerant! Indeed, pollution-favoured

Collectively, the remaining seven species are not a pollution-tolerant assemblage. The high incidence of mosses shows the presence of rock or boulders—large particles not moved in ordinary storms. Whilst this could be "rip-rap" (Fig. 18) or bridges, the other species present indicate that this is a mountain stream.

Fig. 18. "Rip-rap" is a man-made structure of rock or other hard material placed along banks to protect against water-scour. The photo is a bank-reinforced fish-access channel near Byron's Pool, Grantchester, Cambridge and shows medium size stones encased in wire netting (January 2020)

Agrostis stolonifera (Fig. 19) is frequently found at the edges of streams and rivers where storms have eroded and eaten into banks (and clods of earth have fallen into the water) or where banks are low and the grass is on these growing out into the water. Whilst it also occurs in lowlands, it is more characteristic of highlands where there are fewer tall species to shade out this very short grass.

Phalaris arundinacea (Fig. 20) is the tall monocotyledon which extends the furthest up into mountain streams, so, here without, say, *Sparganium erectum*, a mountain stream is likely.

Three mosses, *Agrostis stolonifera* and *Phalaris arundinacea*, all indicate but do not prove that it is a mountain stream, although with all three present together, the probability is high. But since the short edge herbs are the other species present, the **collective** conclusion is: **Mountain stream**.

20

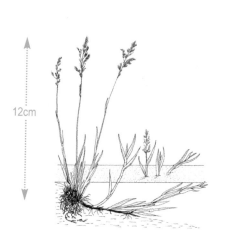

Fig. 19. *Agrostis stolonifera*

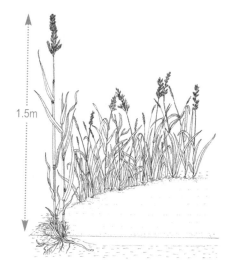

Fig. 20. *Phalaris arundinacea*

When reaching its full size, Butterbur (*Petasites hybridus*, Fig. 21) may occur in streams on soft rocks, but here it is smaller than that. This means that it is growing in a mountain or resistant rock stream (or, of course, in both). This plant is firmly anchored and tolerates much spate flow, so the record could be from any part of the river, upstream or downstream. Downstream, it is probably growing on shoals and shallow edges (if high on the banks its leaves are usually bigger).

Fig. 21. *Petasites hybridus*. Usually grows above normal summer water level but within storm level. Note the comprehensive root-stock to help with anchorage in spates and fast-flowing water

21

With no water-supported species apart from mosses, this river and its bed are unsuitable for larger plants. This could be for various reasons, many of which can be ruled out such as:

- the species composition is wrong for severe or even moderate pollution;

- these species and no others are wrong for a larger or downstream river;

- so many species means there is no continuous heavy shade;

- *Phalaris arundinacea*, *Petasites hybridus* and mosses are not typical of invasion after recent dredging;

- so many edge species show there is no continuous and great disturbance, such as trampling and paddling;

- in very swift, whitewater conditions, fringing herbs would be swept away;

- in deep water, you would expect to see for example, *Ranunculus* or *Myriophyllum* (Water milfoil, *see Fig. 23*).

The most plausible deduction is that the streams are too shallow for water-supported species, apart from mosses. Because of the swift and unstable flow, plants need rather deeper water and more space than they do when growing in lowland streams.

At this type of site, the fringing herbs are at the nutrient-poor end of the range, for example, *Mimulus guttatus* and *Mentha aquatica*. If *Apium nodiflorum* and *Myosotis scorpioides* were present, this would indicate a more nutrient-rich habitat, and add *Berula erecta* for a more calcium-dominant one. However, both these bushy short emergents are easily washed away, so as they are the main species present they may suffer some storm flow, but no great scour and wash-out. This indicates that this site is in the upper, not lower, reaches of the river.

Diagnosis: Deducing from the species list, the habitat is that of a small upper mountain stream, liable to upstream spate only, and on resistant rock or limestone—or very mountainous (alpine) sandstone which carries more vegetation and its silt is more nutrient-rich. The water is shallow, clear, and as nearly-clean as can be found in Britain.

Example 3. Where am I?

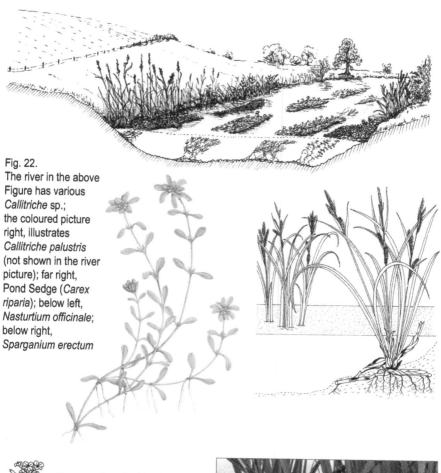

Fig. 22.
The river in the above Figure has various *Callitriche* sp.; the coloured picture right, illustrates *Callitriche palustris* (not shown in the river picture); far right, Pond Sedge (*Carex riparia*); below left, *Nasturtium officinale*; below right, *Sparganium erectum*

Apex leaf is always larger than side leaves

Fig. 22. Continued…/

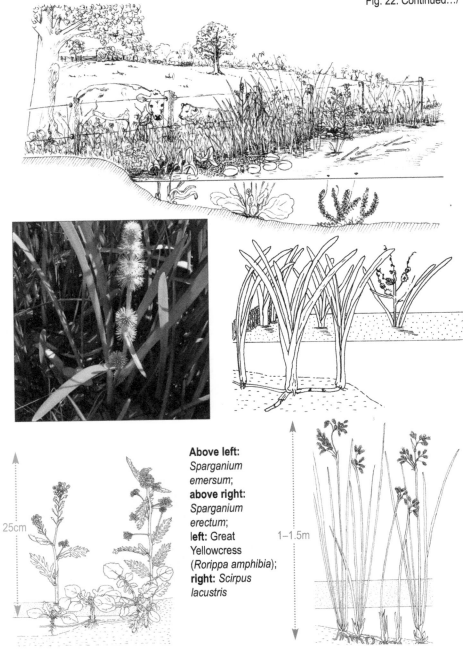

Above left:
Sparganium emersum;
above right:
Sparganium erectum;
left: Great Yellowcress (*Rorippa amphibia*);
right: *Scirpus lacustris*

25cm

1–1.5m

Fig. 22. Continued.../

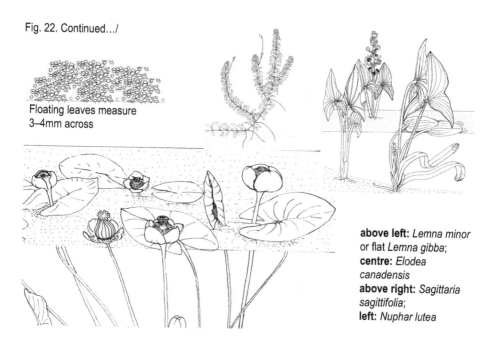

Floating leaves measure
3–4mm across

above left: *Lemna minor*
or flat *Lemna gibba*;
centre: *Elodea canadensis*
above right: *Sagittaria sagittifolia*;
left: *Nuphar lutea*

Imagine a lowland river, a fairly typical British agricultural landscape—plenty of tributaries and no major towns or industry. A look at the geological map shows a mixture of Chalk or other soft limestone, clay, and some sandstone. **Looking at the species lists alone, is it possible to deduce where these sites are** (*excluding for this purpose, sites and their species lists with obvious major damage*)?

Examples of Site-Identification Species

Site 1 (Fig. 23)

SPECIES: *Apium nodiflorum, Myriophyllum spicatum, Nuphar lutea, Phalaris arundinacea, Sagittaria sagittifolia, Sparganium emersum.*

Six species were counted at this British site, comprising 1 tall monocotyledon, 1 fringing herb and 4 water-supported species. As there are only 6 species, in "Good" growing conditions, it is likely that this site is a fairly small, perhaps 4m-wide stream. But with plenty of water—both *Nuphar lutea* and *Sparganium emersum* do better in over 50cm of water—this suggests a nutrient-rich, clay combination.

Deduction: So it is probably a reach of a middle clay stream. Although this is a rather unlikely scenario, the drying of recent decades (1980s–2020) in Britain, notwithstanding February 2020 being the wettest on record, has resulted in such a site being more likely now to have only 4 or 5 emergents and 1 or 2 water-supported species.

Fig. 23

Water milfoil
(*Myriophyllum spicatum*)

Apium nodiflorum

Fig. 23 Continued.../

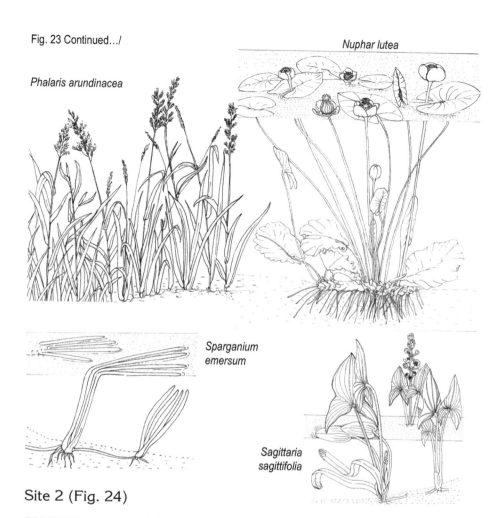

Nuphar lutea

Phalaris arundinacea

Sparganium emersum

Sagittaria sagittifolia

Site 2 (Fig. 24)

SPECIES: *Apium nodiflorum, Berula erecta, Callitriche, Myosotis scorpioides, Ranunculus* species (abundant), *Sparganium erectum.*

As in the **Site 1** example, 6 species are present comprising 1 tall mono-cotyledon, but 3 fringing herbs (counting 2 of these twice) and 4 water-supported species. This undoubtedly points to Soft Limestone judging from the abundant *Ranunculus* and the submerged carpets of *Apium nodiflorum* and *Berula erecta.*

Deduction: The same as **Site 1**. (However, had the site held the same species list, but with no submerged *Apium nodiflorum*, perhaps no *Berula erecta* at

27

all, and the quantities of *Callitriche* and *Ranunculus* species were reversed, the diagnosis would probably have been a sandstone stream.)

Fig. 24

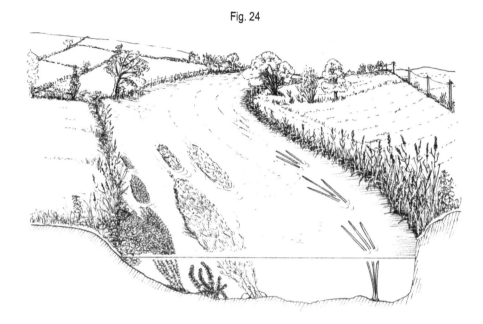

Site 3 (Fig. 25)

SPECIES: *Alisma plantago-aquatica, Elodea canadensis, Enteromorpha intestinalis* (like freshwater seaweed! Fig. 25a), *Lemna minor, Myriophyllum spicatum, Nuphar lutea,* River water-dropwort (*Oenanthe fluviatilis* Fig. 25b), *Ranunculus* spp., *Sagittaria sagittifolia, Scirpus lacustris, Sparganium erectum,* Blanket weed (abundant).

Twelve species! Well! Very high for Lowland England! There are no solely edge species present, but there is *Alisma plantago-aquatica* and *Sparganium erectum* only in shallow water. There is also abundant Blanket weed indicating some domestic or light industrial pollution. Could this be a clay river? *Ranunculus* likes nutrient-medium and flowing water; *Oenanthe fluviatilis* habitat is usually the particular kind of nutrient-medium which comes from mixed limestone (Chalk). It is rather odd that *Sagittaria sagittifolia* is here also, but along with the other species present, shows that the water must be reasonably clean.

Deduction: basically this is a good, at least 8m-wide, large-river vegetation assemblage, growing on clay with a bit of limestone, indicating that there could be other factors such as a Chalk stream entering upstream, or a band of oolite limestone embedded across the catchment.

Fig. 25

Fig. 25a. *Enteromorpha* sp.

Fig. 25b. *Oenanthe fluviatilis* shown here at end of season. Plants are submerged for most of the summer

Site 4 (Fig. 26)

SPECIES: *Agrostis stolonifera*, *Apium nodiflorum*, Horned pondweed (*Zannichellia palustris*, Fig. 26a), *Callitriche* spp., *Glyceria maxima*, Great hairy willowherb or "Codlins and Cream" (*Epilobium hirsutum*, Fig. 26b), *Myosotis scorpioides*, *Myriophyllum spicatum*, *Ranunculus*, *Nasturtium officinale*, *Sparganium erectum*, Blanket weed.

There are twelve species present, but a surprising number do not overlap. One tall monocotyledon, three fringing herbs (none of which have formed submerged carpets), the rest being water-supported species.

Deduction: There are no carpets of *Apium nodiflorum* or *Berula erecta*, indeed there is NO *Berula*! The three less-favouring "Chalky" species are Blanket weed, *Myriophyllum spicatum* and *Zannichellia palustris*. And probably there was a fairly recent minor landslip, in view of the presence of *Agrostis stolonifera* and *Epilobium hirsutum*. This is the most likely way that these (usually) land emergents arrived in this river and, with a strong calcium influence, the species live a surprisingly long time in the water. None of the other species present favour nutrient-rich conditions—so primarily the "rock" at this particular reach in the river must be limestone only, suggesting a large **Chalk stream** river.

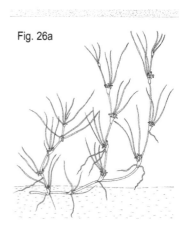

Fig. 26a

Fig. 26

30

Further examples of bioindicator river-types (Fig. 27.1–10)

…And a very healthy sign: if, whilst on your riverine travels, you ever see the "King of Colour" which flies swiftly along many of Britain's waterways—the beautiful bright blue and orange Kingfisher, here sitting amongst reeds (Fig. 27)—it will just make your day!

Fig. 27.
Kingfisher
(*Alcedo atthis*)

Fig. 27.1

Figure 27.1 shows a lop-sided large stream. The banks are diverse and complex, both gentle, and with cliffs. It is a natural shape. Water depth and silting vary and this type of river is "Good" for diversity of plant and animal life.

Plant species include: *Ranunculus fluitans, Phalaris arundinacea, Sparganium emersum* and *Elodea canadensis.*

Fig. 27.2

This wide, upland limestone stream (Fig. 27.2) has a variable bed and plants. There are no boulders, and the river is restricted within the valley.

Plant species include: *Ranunculus* spp., *Mentha aquatica, Myosotis scorpioides* and *Nasturtium officinale* agg.

Fig. 27.3

Figure 27.3 is a Highland mountain stream in resistant rock, eroding to stones, but undredged with natural gravel bars.

Plant species include: *Petasites hybridus* but these, if present, will be small. Very little other vegetation.

Figure 27.4 is a medium size hard rock stream in the Southern Apennines (Italy) with unstable, unconsolidated stone and gravel bed. Banks are over-uniform but okay. The left side has loose stone; the right side overhanging trees. The bed is quite nice to look at, but is really bad for river life. Water Plant species totally absent.

Fig. 27.5

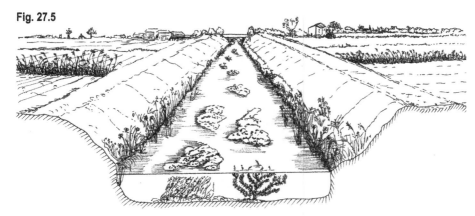

Figure 27.5 shows a medium size, lime-rich, embanked, flowing channel in an alluvial plain north of the Po river (Northern Italy).

Plant species include: *Myriophyllum spicatum, Carex acuta* agg. (Slender-tufted sedge), *Scirpus lacustris, Iris pseudacorus,* Blanket weed and, in the side dykes, Common Reed (*Phragmites australis*).

Fig. 27.6

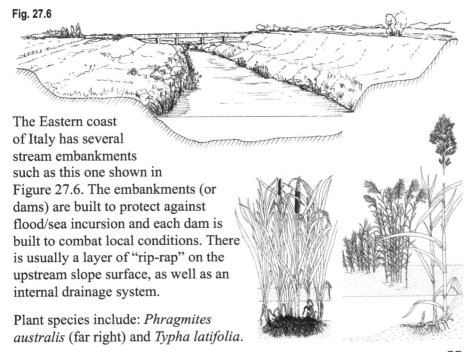

The Eastern coast of Italy has several stream embankments such as this one shown in Figure 27.6. The embankments (or dams) are built to protect against flood/sea incursion and each dam is built to combat local conditions. There is usually a layer of "rip-rap" on the upstream slope surface, as well as an internal drainage system.

Plant species include: *Phragmites australis* (far right) and *Typha latifolia*.

Fig. 27.7a

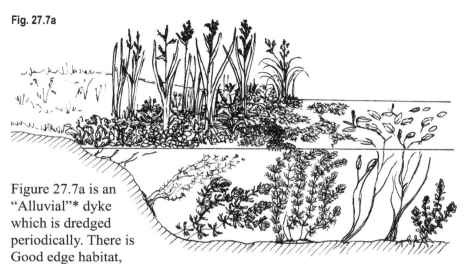

Figure 27.7a is an "Alluvial"* dyke which is dredged periodically. There is Good edge habitat, Good vegetation, emergents on the edge, and water-supported species on the bed, indicating a lime-influence.

Plant species include: *Carex acutiformis, Veronica anagallis-aquatica* agg., *Nasturtium officinale* agg., *Veronica beccabunga, Myosotis scorpioides, Lemna minor* agg., *Callitriche platycarpa, Ceratophyllum demersum, Myriophyllum spicatum, Elodea canadensis, Potamogeton natans.*

Figure 27.7b is also an alluvial dyke, but there is more silt, over-steep banks and Poor emergents. But there is still Good water-supported vegetation.

Fig. 27.7b

Plant species include: *Sagittaria sagittifolia, Callitriche platycarpa, Potamogeton natans, Elodea canadensis, Myriophyllum spicatum, Potamogeton pectinatus.*

---

*"Alluvial". Alluvium refers to deposits of silt, sand, clay, and gravel, and much organic matter. Deposits usually end up in the lower part of a river's course, forming floodplains and deltas. In dykes, however, because of their slow flow, the particles usually settle along the whole length of the waterway—hence the need to dredge periodically to keep the waterway open and flowing well.

Fig. 27.7. Continued.../

Ceratophyllum demersum (Rigid hornwort)

Nasturtium officinale

Carex acutiformis
(Pond-sedge)

Fig. 27.8

The high-quality riverine architecture shown in Figure 27.8 is from a large, lowland, limestone stream in Ireland.

Plant species include: *Ranunculus* spp., *Glyceria maxima,* Jointed rush (*Juncus articulatus*, right)*, Iris pseudacorous, Oenanthe crocata, Sparganium erectum, Phalaris arundinacea* and *Alisma plantago-aquatica.*

Fig. 27.9

This large, upland limestone stream (Fig. 27.9) is rather uniform so has a low species diversity and unchanging bank pattern. It is slightly improved by intermittent trees along the bank edge, but generally not very good ecologically.

Plant species include: *Ranunculus* spp., *Mentha aquatica, Myosotis scorpioides* and *Nasturtium officinale* agg.

Fig. 27.10

Figure 27.10 shows a medium size stream, pollarded willows and lowland open meadows. Plant species at this site, in this state, are negligible.

After disturbance, whether due to storm, animal or human impacts, habitats may open, allowing re-invasion, and then close; ephemeral as well as long-lived species may enter. Even though aquatic vegetation is very low here, the river bed and bank shape will permit recovery, if disturbance is prevented.

Over-little disturbance has less effect, over-much may prevent vegetation growth in the long term. Vegetation can also change from year to year. In the absence of disturbance, where plants grew previously, they will grow again.

BEWARE...!

All the above reads as so easy. Indeed it sometimes is easy, but there are a few pitfalls. Not all plants grow neatly into the central part of their habitat bands, and odd results need to be checked a second or possibly a third time!

(1) Any species, instead of occurring in the middle of its range of flow or substrate, can occur at the extreme end of the range—in negligible rather than moderate flow, for instance. The more species there are, the more likely an anomalous result can be ignored. The anomalous result, though, may be most valuable in interpretation. A local hard bed, such as an ex-ford can, for instance, bear luxuriant *Ranunculus* which is absent on the main (softer) dredged bed.

(2) Most species have many different, not quite variations but, strains which are genetically different. (When these grow best in different habitats, they are called "ecotypes"; if normally different genetically, they are termed "genotypes"—but here we only need to remember that they do vary.) So an odd happening may be an odd plant which has arrived from a place a long way off.

(3) It is a bane of any botanist's imagination that an ecologically unsuitable plant has arrived in the past few weeks and has not yet died or been washed away. This can happen with deliberate, well-meaning gardening when planting out pretty coloured water lilies (or other non-indigenous aquatic species) in the stream running through or by a private garden. Most of these plants will die over their first few months. But if they survive, the results of which are beyond the scope of this little book, they continue not to grow quite like a native species, and a *Flora* or relevant organisation needs to be consulted. (If you are worried that an "Alien has landed" and is causing obstruction or hindrance to the growth of incumbent native species, it would be most helpful if you contact local environment authorities and report it.)

The most probable "unsuitable species" (whether alien or native) are likely to have been washed down from nearer their planted sources by floods or other disturbance. If such fragments end up where they can stay and grow, this is the way plants (with vegetative spread) colonise downstream. But it is not quite that simple. A *Ranunculus* stream may have *Ranunculus* growing all along it, from its source to near the sea. Even though it is all *Ranunculus*, as the river gets bigger and with altering habitat, so the plants may change genetically, the leaves perhaps growing longer downstream. If a particular river is fully investigated, it can be seen that there are not just 2–3 different genetic strains, but many. These may flower at different times, first upstream, then progressively downstream.

In the worst case scenario of plants being washed or planted in, this will include "Invasive Aliens". Most foreign (exotic) plants do not like English conditions and will fade away and die. Just a few aliens, such as Floating pennywort (*Hydrocotyle ranunculoides*) will love it in Britain, settle, stay, grow hugely and smother the indigenous plant populations. (Floating pennywort—such a nice name for a species—by 2010 had grown aggressively in south-east England, and has spread westwards to Wales and into the midlands. In East Anglia (Cambridgeshire) it has caused much disruption to the waterways, Fig. 28a-b.)

Fig. 28a. Grantchester Mill Race in September 2017. "Before" and "After" of mechanical removal of floating pennywort in the River Cam. (Photograph, courtesy of the Environment Agency ©)

Fig. 28b. Another "Before" and "After" for floating pennywort—Bottisham Lock, River Cam, November 2017 (photograph, courtesy of Mike Foley ©). (Published by the Cam Valley Forum ©, *Cam Valley Matters*, No. 42. 1, December 2017 Newsletter)

Another invasive plant to Great Britain: Himalayan balsam (*Impatiens glandulifera*)

Distribution of *Apium nodiflorum* (Fool's water-cress) and *Berula erecta* (Lesser water parsnip) in rivers in Western Europe

Overview:

Why make a "case" of these two particular aquatic plants? Mainly because both plants are quite common so therefore invaluable as "bioindicators"— they are native to Britain and their distribution has been recorded since the 1930s. They are also interesting in that their similarities when young make identification of both plants almost impossible for the novice: even the expert eye needs a leaf in hand to separate underwater populations! The *Online Atlas of the British and Irish Flora* lists both species as being present in the 1962 published *Atlas*. The 2020 "status" of *Berula erecta* in Britain is: "…Declines have occurred, and analysis of the database reveals that most of these losses have occurred since 1950. They are probably caused by drainage and habitat destruction." *Apium nodiflorum*, with its wider habitat range, seems to have fared better: "…The distribution of this species is stable."

We have already "deduced" that each river plant has its own (sometimes variable) form, habitat and individual function in that habitat. Of the short and bushy aquatic plants, however, only *Apium nodiflorum* and *Berula erecta* grow in submerged carpets in shallow water! The habitats of both species significantly overlap and they live and grow closely together in mixed communities. The young leaves of submerged *Berula erecta* and *Apium nodiflorum* look amazingly similar. Scrutiny of a submerged carpet of plants does not readily show that there are two species growing together—they are virtually indistinguishable. But, for example, later in the year when the plants come into flower it is easy to see that the white, umbelled *Apium nodiflorum* flowers are tucked modestly amongst the plant's leaves on short stalks (in the "armpits"), whereas *Berula erecta* flowers, also white umbels, prominently stretch out above the plant. In fertile situations, *Apium nodiflorum* grows much larger and is a sprawler, reaching up to 2m. If present at all by that stage (not smothered by its neighbour), the more fine *Berula erecta* is probably hardly over 0.75m and does not sprawl. Both species occur sparsely in wetlands or pond edges in a wider area. Table 2 lists both the similarities and the differences where these two plants grow in Western Europe.

Table 2.

Distribution of *Apium nodiflorum* and *Berula erecta* in rivers in Western Europe

GENERAL DISTRIBUTION

(a) Regions without (much):

Apium nodiflorum	Denmark, North Germany
Berula erecta	Central Italy, South Italy, Malta

(b) The two species usually occurring:

Separately	Germany*, Italy, The Netherlands
Together (with more *A. nodiflorum*)	France, Ireland
Both patterns	Britain

DISTRIBUTION IN RELATION TO LIME	*Strong*	*Moderate*	*Weak*	*Negligible*
Apium nodiflorum		North	Britain	South France
		France	Ireland	(Netherlands)
		Germany*	Italy	
Berula erecta	Italy	Britain	C Germany	Denmark
		France		N Germany
		S Germany		
		Ireland		

* The former West Germany

Identification Notes:

Figure 29 shows, schematically, subtle differences between the two species: flowers, root system and leaf stems.

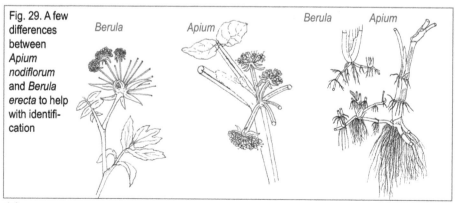

Fig. 29. A few differences between *Apium nodiflorum* and *Berula erecta* to help with identification

Berula   Apium   Berula   Apium

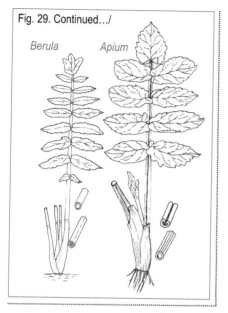

Fig. 29. Continued.../

Berula   Apium

Berula erecta (Lesser water parsnip) (Fig. 30a-b)

General Habit: Native of Western Europe. Grows upstream in rivers and streams and around springs. Winter growth may form a submerged mat; summer growth is mostly above water (emergent) from May. It likes shallow water with moderate nutrient levels, favouring calcareous (Limestone/Chalk) environments. It does not tolerate pollution. English common names include Lesser water-parsnip, Cutleaf water-parsnip and Narrow-leaved water-parsnip. It is a member of the Carrot (*Apiaceae*) family. Stoloniferous perennial—stolons are lateral stems, growing horizontally at ground level from which new plants grow and root at intervals, a bit like strawberry plants; perennial plants live for a number of years. A hairless, loosely-compact plant which grows to about 1m in height. Usually grows all year.

Description:

Flowers: Umbels (flowerheads) 3–6cm across, sprout opposite a leaf at the top of a stem. Each umbel can branch into 10 or 20 slim, smooth "rays" (spokes, like an umbrella). Each umbel also has a set of large, leaf-like bracts at its base as well as a smaller set (bracteoles) at the base of the flower head on each ray. The tiny individual flowers appear July to September, are white, about 2mm across with 5 petals, a creamy white centre, and 5 thin, white stamens.

Fig. 30a. *Berula erecta* identification, line art sketch showing growth pattern

Leaves: Pinnate (along a stalk). Aerial leaves consist of 7–14 pairs of leaflets measuring 2–6cm. The individual leaflets are toothed (serrate), lanceolate (spear-head shape), with the uppermost ones gradually becoming smaller. Submerged leaves have 3–4 pairs of leaflets with long, narrow parallel lobes (linear).

Stems: Hollow (fistular) and lined, with a pale ring-mark (septum—more visible if the leaf is held up to the light) at the base of the stem.

Roots: Fibrous mass of white roots of differing lengths and widths, and connecting stolons.

Apium nodiflorum (Fool's water cress, synonym: *Helosciadium nodiflorum*) (Fig. 31a-b)

General Habit: Native of Western Europe. Grows in shallow water in streams and lakes, also in ponds, ditches, canals, marshes, and on intermittent-flood ground in fine silt or mud. It favours nutrient-rich shallow water. English common names include Fool's water cress, Lebanese cress, Poor man's watercress. It is a member of the Carrot (*Apiaceae*) family. Prostrate to ascending perennial—low-growing hairless "scrambler" (but climbs and can smother other plants, including *Berula erecta*); perennial plants live for a number of years. Grows to about 60cm in height, but can "scramble" to 2m. Usually grows all year, submerged in winter months.

Description:

Flowers: Umbels. Flowers are small and white with 5 petals, and one petal can be slightly longer than the rest. They sit at the top of 3–12 (thickish) finely-grooved rays. Umbels measure 3–5cm across, grow opposite a leaf from a short stalk or right against the stem at a leaf axil ("armpit"). There are no bracts at the stalk juncture with the stem (unlike *Berula erecta*), but each individual flowerhead has 4–7 small bracteoles circling beneath. Blooms in July and August.

Leaves: Complex leaf with 4–6 pairs of opposite leaflets (lanceolate or rounded) with rounded or pointed tooth margins along a central leaf stalk with a single leaflet at the top.

Fig. 30b. *Berula erecta*. Colour plate shows umbel bud, flowering umbels and seedhead

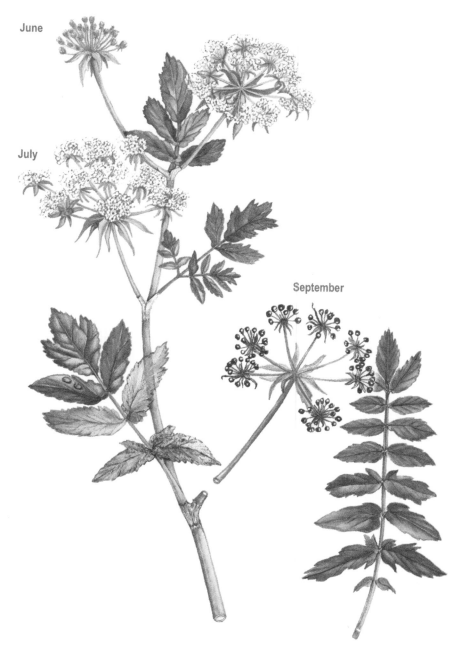

June

July

September

Stems: Upper stems Hollow. Finely grooved, rooting at the lower nodes. Variously angular or ridged, or both.

Roots: Many fine white roots of varying lengths. Also sprout at lower-leaf-stalk intersections. Sprawling stolons with large terminal shoots. Roots also occur at intervals along the stolons.

Fig. 31a. *Apium nodiflorum* identification, line art sketch showing growth pattern

Fig. 31b. *Apium nodiflorum*.
Colour plate shows flowering
umbels and seedhead

June

August

49

Deduction:

As you can see, it is quite a task to separate these two plants. Figure 32a-d shows both plants growing in lime-base streams in Cambridgeshire. Figure 32a shows a submerged carpet in February and Figure 32b the same site in May (note the bed is dry!).

Fig. 32a. Submerged carpet (February) and b, the emergent vegetation growing in the same place in May (2020—dry bed)

The patches of vegetation shown in Figures 32c-d are growing in the same brook, photographed in May. The physical distance between the two pictures is about 8m—it is still almost impossible to tell which plant is which! Perhaps a tiny fringe of *Berula erecta* is at the back, and *Apium nodiflorum* is mid-stream and forefront in both photographs.

Fig. 32c-d. *Apium nodiflorum* (middle and foreground) and *Berula erecta* (back fringe) growing together in a small lime-based stream (Tit Brook) in Cambridgeshire in May 2020

Figure 33a-b shows the Western European [surveyed] river distribution of *Berula erecta* and *Apium nodiflorum* (1970s–80s). How odd! There is no very consistent pattern! Figure 33c amalgamates the distribution maps of 33a-b and shows that these two species can occur together, in the same river (or at least in the same habitat in that river).

They can also be different. How come? It is plausible to assume that geographic factors stop *Berula erecta* growing in the south—hotter summers, hotter winters, dry mid and late summers—or any or all of these (or just possibly the difference in other climatic factors such as light). That would be difficult enough to work out, but why should *Apium nodiflorum* be more frequent in the west?

Then, when the two occur in the same or in nearby rivers why, in some countries or places, should the two occur separately whilst in others they also occur together? Just to say "Oh well, the climate is different" begs the question, since the two are commonly together in Britain, France and Ireland, but not in The Netherlands or Italy (both adjacent to France). Management perhaps?

Finally, what about calcium or lime? In Italy, for instance, if you think you can see *Apium nodiflorum* or *Berula erecta* on lime and go towards it knowing confidently that you would expect to see *Berula—Berula erecta* it is. Do the same in Denmark or Northern Germany, and it is just as likely to be *Apium nodiflorum* on sands or sandstone! And not just because these are northern sites, since the same would also be true in Southern France.

Fig. 33a
Distribution Key of river sites surveyed
bearing *Apium nodiflorum*

☐	0–4%
▒	5–14%
⋯	15–24%
⊡	25–49%
■	50+%

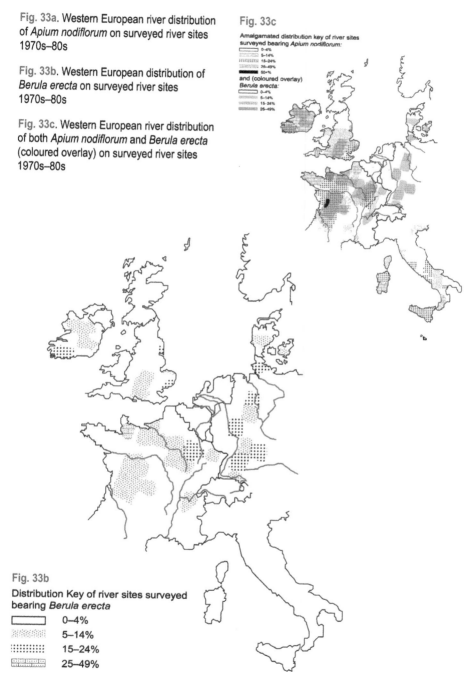

Fig. 33a. Western European river distribution of *Apium nodiflorum* on surveyed river sites 1970s–80s

Fig. 33b. Western European distribution of *Berula erecta* on surveyed river sites 1970s–80s

Fig. 33c. Western European river distribution of both *Apium nodiflorum* and *Berula erecta* (coloured overlay) on surveyed river sites 1970s–80s

Fig. 33c

Amalgamated distribution key of river sites surveyed bearing *Apium nodiflorum*:
- 0–4%
- 5–14%
- 15–24%
- 25–49%
- 50%

and (coloured overlay)
Berula erecta:
- 0–4%
- 5–14%
- 15–24%
- 25–49%

Fig. 33b

Distribution Key of river sites surveyed bearing *Berula erecta*
- 0–4%
- 5–14%
- 15–24%
- 25–49%

53

Synopsis

Diagnostic deductions do come but not just by, for example, the presence of a specific Chalk species. A larger river receives more "run-off" (influenced by fertiliser, road run-off, or sewage treatment works, and other effluents) so in some reaches there will be more non-Chalk influence.

River plants are excellent bioindicators, expressing conditions of the life, health and structure of a river. They are mostly large enough to be seen from the river bank, a boat, or a bridge. There is usually no need for trawling (manually with a weighted grappling hook on a piece of string), or identification through a magnifying glass or a microscope. The common species are of suitable number to be easily learnt. Using just 70 species, the common British river habitats can all be identified and assessed. Using 120 species refines the process and identifies all European river types also. It is relatively easy to learn 100 species names, whether English or *Latin*!

Moreover, just *Sparganium erectum* present means any non-torrent, non-bog, non-grossly polluted or disturbed stream. Add much *Lemna minor*, and flow is nil to moderate. Add *Nuphar lutea* and water reaches at least 0.5m deep. Add *Sagittaria sagittifolia* and disturbance and pollution are not great. Add *Apium nodiflorum* and there is some sloping, silty or earthen bank. Add *Myosotis scorpioides* and nutrients are plentiful, and so on.

River plants have the advantage of being stationary. They do not swim or run away from observers—but can be uprooted and disappear after a spate, or heavy flooding. They are extremely sensitive to the habitats in which they live and therefore, their presence, pattern, amount and behaviour can all help to interpret their habitat conditions. As site species number increases, so does the accuracy of the diagnosis; the habitat range in which all the site species can occur becoming increasingly restricted.

Truly, anyone can learn the species of a small area quite quickly, probably only *c.* 25 species (if you are lucky enough to come across a rare species, it can usually be ignored for the purposes described here).

The species lists and diagnoses are endless: three more plants (which grow in bogs, marshes, swamps, dunes, and woodland areas beside streams—aquatic areas not covered by this book) are shown in Figure 34a–c. So hopefully you can now see that **the plants do indeed speak**?

Top right: Fig. 34a. *Hottonia palustris* (Water violet) is naturally a **bog or marsh plant**, with stems reaching up to 80cm in height. Its basal roots are buried in the underlying mud, while other silvery, shiny roots dangle freely in the water. In watercourses usually in dykes or drains

Bottom right: Fig. 34b. *Menyanthes trifoliata* (Bog bean) grows as an emergent at the shallow edge of lakes, pools or **slow-flowing rivers**, swamps, bog streams, flushes or dune-slacks. It tolerates a wide range of water chemistry, but not shade. In water courses, usually in low nutrients (oligotrophic)

Bottom left: Fig. 34c. *Butomus umbellatus* is a British lowland native plant with lovely pink flower umbels. It grows to 1m in height as an emergent at the edges of rivers, lakes, canals, ditches and in swamps. In watercourses, it usually grows in fertile mud and water up to 25cm deep, in full sunlight.

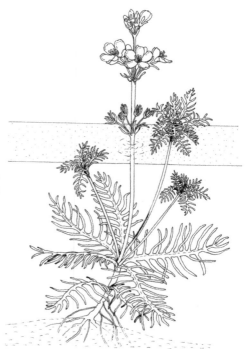

Conclusion

We know an enormous amount about river plants. But also, there is so much more we do not know! Beware, indeed, of transferring "logical" conclusions from one region to another. Most of the time it is safe to do so, and all of the time it is safe to do so within small areas which have already been recorded and whose communities and habitats are known. By using a species list and Classification System the right conclusions can easily be drawn. Note: it is unlikely that a dozen species are all behaving oddly, so the data is significant enough to warrant further investigation. As already pointed out, though, the larger the number of species the observer is working with, the more accurate will be the diagnosis.

Figure 35 shows a beautiful Chalk river with lots of vegetation. There are about 12 species present, including *Callitriche* spp., *Ranunculus* spp. (*Ranunculus calcareous*, *Ranunculus trichophyllus*), *Veronica anagallis-aquatica* agg., *Mentha aquatica*, *Nasturtium officinale* agg., *Myosotis scorpioides*, *Berula erecta* (white flowers, bottom centre), *Sparganium erectum*, *Phalaris arundinacea*, *Glyceria maxima*, *Veronica beccabunga*, and *Agrostis stolonifera*.

Fig. 35. Some aquatic vegetation in a Chalk stream (River Itchen, Hampshire, August 2011)

REFERENCES

Cam Valley Forum, *Cam Valley Matters*, No. 42. 1 December 2017 Newsletter.

Clapham, A. R.; Tutin, T. G.; & Warburg, E. F. (1952). *Flora of the British Isles*. Cambridge University Press.

Davies, W.H. (1911). "Leisure" is a poem by Welsh poet William Henry Davies, appearing originally in his *Songs Of Joy and Others*, published in 1911 by A. C. Fifield, and then in Davies' first anthology *Collected Poems* by the same publisher in 1916. (Source: Wikipedia.) This Poem is now in the Public Domain so copyright (©) does not apply.

Holmes, N. (1983) *Typing British Rivers according to their Flora*. Focus on Nature Conservation, No. 4, Peterborough. Nature Conservancy Council.

Table 1: Modified from Haslam, S.M. & Wolseley, P.A. (1981). *River Vegetation: Its Identification, Assessment And Management*. Cambridge University Press, Cambridge.

The Online Atlas of the British and Irish Flora. Biological Records Centre (BRC). URL https://www.brc.ac.uk/plantatlas/. Accessed 11 June 2020.

THE RIVER FRIEND SERIES

This series of small books is designed for people with a general or specific interest in rivers.
Please visit the River Friend Website for an up to-date list of
PUBLISHED Titles: **http://www.riverfriend.tinasfineart.uk /home**

Standalone Titles in the Series include:

A PROLOGUE TO THE SERIES: Plant identification and Glossary of Terms
(ISBN 978 1 9162096 2 6)

DRYING UP (ISBN 978 1 9162096 1 9)

STREAM STORY I: A Riveting Riverscape—River Brue, Somerset
(ISBN 978 1 9162096 0 2)

INTERPRET: What do Plants Tell us? (ISBN 978 1 9162096 5 7)

Vegetation Changes Over Time. Is there FREEZE FRAME?
(ISBN 978 1 9162096 6 4)

REED—ON THE EDGE (ISBN 978 1 9162096 4 0)

An Introduction to the WATER FRAMEWORK DIRECTIVE
(ISBN 978 1 9162096 3 3)

WATER: Clean and Dirty (ISBN 978 1 9162096 7 1)

STREAM STORY II: A Brook in Transit: Bourn Brook, Cambs
(ISBN 978 1 9162096 8 8)

CHANGE: What a Disaster! (ISBN 978 1 9162096 9 5)

LOOK AT THE BOTTOM

How to lose Fresh Water in Under Two Centuries. The Example of MALTA

VEGETATION PATTERNS

IN THE WATER

THE WATERS OF WELLS

RESTORE, REHABILITATE, IMPROVE

AWFUL ALIENS

About the Authors

Sylvia Haslam is a botanist and river culture, etc., specialist. Anyone wanting to find out more should look at the publications list on her website (http://www.riversandreeds.co.uk). Her publications specific to this series are listed in the book entitled *A PROLOGUE TO THE SERIES: Plant identification and Glossary of Terms*.

Tina Bone has worked as a self-employed Desktop Publisher for many years until she changed career to work as a Professional Artist from March 2005. To view Tina's resumé and artwork please visit her website: http://www.tinasfineart.uk.

Lightning Source UK Ltd.
Milton Keynes UK
UKHW021425210620
365274UK00007B/72